TABLE OF CONTENTS

S0-CDP-373

STUDENT NAME: ALZBETA MOLNAROVA   ADDRESS: 45 MOUNT ROYAL DR   TELEPHONE NUMBER: 778 833 0995

NETWORK ID: 300273153   COURSE: _____   SECTION: _____

SEMESTER: _____   LABORATORY INSTRUCTOR: _____

LABORATORY PARTNERS: _____

DATE	EXPERIMENT/SUBJECT	PAGE NO.

TABLE OF CONTENTS *(continued)*

Notebook No: _____

DATE	EXPERIMENT/SUBJECT	PAGE NO.

EXP. NUMBER	EXPERIMENT/SUBJECT		DATE	
NAME		LAB PARTNER	LOCKER/DESK NO.	COURSE & SECTION NO.

SIGNATURE	DATE	WITNESS/TA		DATE

EXP. NUMBER	EXPERIMENT/SUBJECT		DATE	
NAME		LAB PARTNER	LOCKER/DESK NO.	COURSE & SECTION NO.

SIGNATURE		DATE	WITNESS/TA		DATE

EXP. NUMBER	EXPERIMENT/SUBJECT		DATE	
NAME		LAB PARTNER	LOCKER/DESK NO.	COURSE & SECTION NO.

02

SIGNATURE	DATE	WITNESS/TA		DATE

THE HAYDEN-McNEIL STUDENT LAB NOTEBOOK   NOTE: INSERT DIVIDER UNDER COPY SHEET BEFORE WRITING

EXP. NUMBER	EXPERIMENT/SUBJECT		DATE	
NAME		LAB PARTNER	LOCKER/DESK NO.	COURSE & SECTION NO.

EXP. NUMBER	EXPERIMENT/SUBJECT		DATE	
SIGNATURE		DATE	WITNESS/TA	DATE

THE HAYDEN-McNEIL STUDENT LAB NOTEBOOK

NOTE: INSERT DIVIDER UNDER COPY SHEET BEFORE WRITING

EXP. NUMBER	EXPERIMENT/SUBJECT		DATE	
NAME		LAB PARTNER	LOCKER/DESK NO.	COURSE & SECTION NO.

SIGNATURE		DATE	WITNESS/TA		DATE

NOTE: INSERT DIVIDER UNDER COPY SHEET BEFORE WRITING

EXP. NUMBER	EXPERIMENT/SUBJECT		DATE	
NAME		LAB PARTNER	LOCKER/DESK NO.	COURSE & SECTION NO.

EXP. NUMBER	EXPERIMENT/SUBJECT			
SIGNATURE		DATE	WITNESS/TA	DATE

NOTE: INSERT DIVIDER UNDER COPY SHEET BEFORE WRITING

EXP. NUMBER	EXPERIMENT/SUBJECT		DATE	
NAME		LAB PARTNER	LOCKER/DESK NO.	COURSE & SECTION NO.

SIGNATURE		DATE	WITNESS/TA		DATE

THE HAYDEN-McNEIL STUDENT LAB NOTEBOOK

EXP. NUMBER	EXPERIMENT/SUBJECT		DATE	
NAME		LAB PARTNER	LOCKER/DESK NO.	COURSE & SECTION NO.

SIGNATURE	DATE	WITNESS/TA	DATE

05

| EXP. NUMBER | EXPERIMENT/SUBJECT | | DATE | |
| NAME | | LAB PARTNER | LOCKER/DESK NO. | COURSE & SECTION NO. |

| SIGNATURE | DATE | WITNESS/TA | DATE |

THE HAYDEN-McNEIL STUDENT LAB NOTEBOOK   NOTE: INSERT DIVIDER UNDER COPY SHEET BEFORE WRITING

EXP. NUMBER	EXPERIMENT/SUBJECT		DATE	
NAME		LAB PARTNER	LOCKER/DESK NO.	COURSE & SECTION NO.

05

COPY

EXP. NUMBER	EXPERIMENT/SUBJECT		DATE	

SIGNATURE	DATE	WITNESS/TA	DATE

THE HAYDEN-McNEIL STUDENT LAB NOTEBOOK

NOTE: INSERT DIVIDER UNDER COPY SHEET BEFORE WRITING

EXP. NUMBER	EXPERIMENT/SUBJECT		DATE	
NAME		LAB PARTNER	LOCKER/DESK NO.	COURSE & SECTION NO.

SIGNATURE	DATE	WITNESS/TA	DATE

THE HAYDEN-McNEIL STUDENT LAB NOTEBOOK

NOTE: INSERT DIVIDER UNDER COPY SHEET BEFORE WRITING

EXP. NUMBER	EXPERIMENT/SUBJECT		DATE	
NAME		LAB PARTNER	LOCKER/DESK NO.	COURSE & SECTION NO.

EXP. NUMBER	EXPERIMENT/SUBJECT		DATE	
SIGNATURE		DATE	WITNESS/TA	DATE

NOTE: INSERT DIVIDER UNDER COPY SHEET BEFORE WRITING

EXP. NUMBER	EXPERIMENT/SUBJECT		DATE	
NAME		LAB PARTNER	LOCKER/DESK NO.	COURSE & SECTION NO.

SIGNATURE	DATE	WITNESS/TA		DATE

THE HAYDEN-McNEIL STUDENT LAB NOTEBOOK

EXP. NUMBER	EXPERIMENT/SUBJECT		DATE	
NAME		LAB PARTNER	LOCKER/DESK NO.	COURSE & SECTION NO.

SIGNATURE	DATE	WITNESS/TA	DATE

EXP. NUMBER	EXPERIMENT/SUBJECT		DATE	
NAME		LAB PARTNER	LOCKER/DESK NO.	COURSE & SECTION NO.

SIGNATURE	DATE	WITNESS/TA	DATE

THE HAYDEN-McNEIL STUDENT LAB NOTEBOOK　　　NOTE: INSERT DIVIDER UNDER COPY SHEET BEFORE WRITING

EXP. NUMBER	EXPERIMENT/SUBJECT		DATE	
NAME		LAB PARTNER	LOCKER/DESK NO.	COURSE & SECTION NO.

SIGNATURE		DATE	WITNESS/TA	DATE

EXP. NUMBER	EXPERIMENT/SUBJECT		DATE	
NAME		LAB PARTNER	LOCKER/DESK NO.	COURSE & SECTION NO.

SIGNATURE		DATE	WITNESS/TA		DATE

THE HAYDEN-McNEIL STUDENT LAB NOTEBOOK

NOTE: INSERT DIVIDER UNDER COPY SHEET BEFORE WRITING

EXP. NUMBER	EXPERIMENT/SUBJECT		DATE	
NAME		LAB PARTNER	LOCKER/DESK NO.	COURSE & SECTION NO.

SIGNATURE		DATE	WITNESS/TA		DATE

EXP. NUMBER	EXPERIMENT/SUBJECT		DATE	
NAME		LAB PARTNER	LOCKER/DESK NO.	COURSE & SECTION NO.

SIGNATURE		DATE	WITNESS/TA		DATE

EXP. NUMBER	EXPERIMENT/SUBJECT		DATE	
NAME		LAB PARTNER	LOCKER/DESK NO.	COURSE & SECTION NO.

SIGNATURE	DATE	WITNESS/TA	DATE

NOTE: INSERT DIVIDER UNDER COPY SHEET BEFORE WRITING

EXP. NUMBER	EXPERIMENT/SUBJECT		DATE	
NAME		LAB PARTNER	LOCKER/DESK NO.	COURSE & SECTION NO.

SIGNATURE		DATE	WITNESS/TA		DATE

THE HAYDEN-McNEIL STUDENT LAB NOTEBOOK

NOTE: INSERT DIVIDER UNDER COPY SHEET BEFORE WRITING

EXP. NUMBER	EXPERIMENT/SUBJECT		DATE	
NAME		LAB PARTNER	LOCKER/DESK NO.	COURSE & SECTION NO.

EXP. NUMBER	EXPERIMENT/SUBJECT		DATE	
SIGNATURE		DATE	WITNESS/TA	DATE

EXP. NUMBER	EXPERIMENT/SUBJECT		DATE	
NAME		LAB PARTNER	LOCKER/DESK NO.	COURSE & SECTION NO.

SIGNATURE		DATE	WITNESS/TA		DATE

NOTE: INSERT DIVIDER UNDER COPY SHEET BEFORE WRITING

EXP. NUMBER	EXPERIMENT/SUBJECT		DATE	
NAME	LAB PARTNER		LOCKER/DESK NO.	COURSE & SECTION NO.

SIGNATURE	DATE	WITNESS/TA		DATE

NOTE: INSERT DIVIDER UNDER COPY SHEET BEFORE WRITING

EXP. NUMBER	EXPERIMENT/SUBJECT		DATE	
NAME		LAB PARTNER	LOCKER/DESK NO.	COURSE & SECTION NO.

SIGNATURE	DATE	WITNESS/TA	DATE

THE HAYDEN-McNEIL STUDENT LAB NOTEBOOK

EXP. NUMBER	EXPERIMENT/SUBJECT		DATE	
NAME		LAB PARTNER	LOCKER/DESK NO.	COURSE & SECTION NO.

SIGNATURE		DATE	WITNESS/TA		DATE

THE HAYDEN-McNEIL STUDENT LAB NOTEBOOK

EXP. NUMBER	EXPERIMENT/SUBJECT		DATE	
NAME		LAB PARTNER	LOCKER/DESK NO.	COURSE & SECTION NO.

SIGNATURE		DATE	WITNESS/TA		DATE

NOTE: INSERT DIVIDER UNDER COPY SHEET BEFORE WRITING

EXP. NUMBER	EXPERIMENT/SUBJECT		DATE	
NAME		LAB PARTNER	LOCKER/DESK NO.	COURSE & SECTION NO.

SIGNATURE		DATE	WITNESS/TA		DATE

NOTE: INSERT DIVIDER UNDER COPY SHEET BEFORE WRITING

EXP. NUMBER	EXPERIMENT/SUBJECT		DATE	
NAME		LAB PARTNER	LOCKER/DESK NO.	COURSE & SECTION NO.

SIGNATURE	DATE	WITNESS/TA		DATE

THE HAYDEN-McNEIL STUDENT LAB NOTEBOOK

NOTE: INSERT DIVIDER UNDER COPY SHEET BEFORE WRITING

EXP. NUMBER	EXPERIMENT/SUBJECT		DATE	
NAME		LAB PARTNER	LOCKER/DESK NO.	COURSE & SECTION NO.

SIGNATURE		DATE	WITNESS/TA		DATE

EXP. NUMBER	EXPERIMENT/SUBJECT		DATE	
NAME		LAB PARTNER	LOCKER/DESK NO.	COURSE & SECTION NO.

SIGNATURE		DATE	WITNESS/TA		DATE

EXP. NUMBER	EXPERIMENT/SUBJECT		DATE	
NAME		LAB PARTNER	LOCKER/DESK NO.	COURSE & SECTION NO.

SIGNATURE		DATE	WITNESS/TA		DATE

NOTE: INSERT DIVIDER UNDER COPY SHEET BEFORE WRITING

EXP. NUMBER	EXPERIMENT/SUBJECT		DATE	
NAME		LAB PARTNER	LOCKER/DESK NO.	COURSE & SECTION NO.

SIGNATURE	DATE	WITNESS/TA	DATE

NOTE: INSERT DIVIDER UNDER COPY SHEET BEFORE WRITING

EXP. NUMBER	EXPERIMENT/SUBJECT		DATE	
NAME		LAB PARTNER	LOCKER/DESK NO.	COURSE & SECTION NO.

COPY

SIGNATURE		DATE	WITNESS/TA	DATE

NOTE: INSERT DIVIDER UNDER COPY SHEET BEFORE WRITING

EXP. NUMBER	EXPERIMENT/SUBJECT		DATE	
NAME		LAB PARTNER	LOCKER/DESK NO.	COURSE & SECTION NO.

SIGNATURE		DATE	WITNESS/TA		DATE

EXP. NUMBER	EXPERIMENT/SUBJECT		DATE	
NAME		LAB PARTNER	LOCKER/DESK NO.	COURSE & SECTION NO.

COPY

SIGNATURE		DATE	WITNESS/TA		DATE

NOTE: INSERT DIVIDER UNDER COPY SHEET BEFORE WRITING

EXP. NUMBER	EXPERIMENT/SUBJECT		DATE	
NAME		LAB PARTNER	LOCKER/DESK NO.	COURSE & SECTION NO.

SIGNATURE		DATE	WITNESS/TA		DATE

EXP. NUMBER	EXPERIMENT/SUBJECT		DATE	
NAME		LAB PARTNER	LOCKER/DESK NO.	COURSE & SECTION NO.

SIGNATURE	DATE	WITNESS/TA	DATE

20

EXP. NUMBER	EXPERIMENT/SUBJECT		DATE	
NAME		LAB PARTNER	LOCKER/DESK NO.	COURSE & SECTION NO.

SIGNATURE	DATE	WITNESS/TA	DATE

THE HAYDEN-McNEIL STUDENT LAB NOTEBOOK

EXP. NUMBER	EXPERIMENT/SUBJECT		DATE	
NAME		LAB PARTNER	LOCKER/DESK NO.	COURSE & SECTION NO.

SIGNATURE	DATE	WITNESS/TA	DATE

EXP. NUMBER	EXPERIMENT/SUBJECT		DATE	
NAME		LAB PARTNER	LOCKER/DESK NO.	COURSE & SECTION NO.

SIGNATURE		DATE	WITNESS/TA	DATE

EXP. NUMBER	EXPERIMENT/SUBJECT		DATE	
NAME		LAB PARTNER	LOCKER/DESK NO.	COURSE & SECTION NO.

COPY

SIGNATURE		DATE	WITNESS/TA		DATE

NOTE: INSERT DIVIDER UNDER COPY SHEET BEFORE WRITING

EXP. NUMBER	EXPERIMENT/SUBJECT		DATE	
NAME		LAB PARTNER	LOCKER/DESK NO.	COURSE & SECTION NO.

SIGNATURE		DATE	WITNESS/TA		DATE

EXP. NUMBER	EXPERIMENT/SUBJECT		DATE	
NAME		LAB PARTNER	LOCKER/DESK NO.	COURSE & SECTION NO.

SIGNATURE		DATE	WITNESS/TA		DATE

NOTE: INSERT DIVIDER UNDER COPY SHEET BEFORE WRITING

EXP. NUMBER	EXPERIMENT/SUBJECT		DATE	
NAME		LAB PARTNER	LOCKER/DESK NO.	COURSE & SECTION NO.

SIGNATURE	DATE	WITNESS/TA	DATE

EXP. NUMBER	EXPERIMENT/SUBJECT		DATE	
NAME		LAB PARTNER	LOCKER/DESK NO.	COURSE & SECTION NO.

EXP. NUMBER	EXPERIMENT/SUBJECT		DATE	
SIGNATURE		DATE	WITNESS/TA	DATE

NOTE: INSERT DIVIDER UNDER COPY SHEET BEFORE WRITING

EXP. NUMBER	EXPERIMENT/SUBJECT		DATE	
NAME		LAB PARTNER	LOCKER/DESK NO.	COURSE & SECTION NO.

SIGNATURE		DATE	WITNESS/TA		DATE

EXP. NUMBER	EXPERIMENT/SUBJECT		DATE	
NAME		LAB PARTNER	LOCKER/DESK NO.	COURSE & SECTION NO.

SIGNATURE		DATE	WITNESS/TA		DATE

EXP. NUMBER	EXPERIMENT/SUBJECT		DATE	
NAME		LAB PARTNER	LOCKER/DESK NO.	COURSE & SECTION NO.

SIGNATURE		DATE	WITNESS/TA		DATE

EXP. NUMBER	EXPERIMENT/SUBJECT		DATE	
NAME		LAB PARTNER	LOCKER/DESK NO.	COURSE & SECTION NO.

SIGNATURE		DATE	WITNESS/TA		DATE

NOTE: INSERT DIVIDER UNDER COPY SHEET BEFORE WRITING

EXP. NUMBER	EXPERIMENT/SUBJECT		DATE	
NAME		LAB PARTNER	LOCKER/DESK NO.	COURSE & SECTION NO.

SIGNATURE		DATE	WITNESS/TA		DATE

EXP. NUMBER	EXPERIMENT/SUBJECT		DATE	
NAME		LAB PARTNER	LOCKER/DESK NO.	COURSE & SECTION NO.

SIGNATURE	DATE	WITNESS/TA		DATE

THE HAYDEN-McNEIL STUDENT LAB NOTEBOOK

NOTE: INSERT DIVIDER UNDER COPY SHEET BEFORE WRITING

EXP. NUMBER	EXPERIMENT/SUBJECT		DATE	
NAME		LAB PARTNER	LOCKER/DESK NO.	COURSE & SECTION NO.

SIGNATURE		DATE	WITNESS/TA		DATE

EXP. NUMBER	EXPERIMENT/SUBJECT		DATE	
NAME		LAB PARTNER	LOCKER/DESK NO.	COURSE & SECTION NO.

SIGNATURE	DATE	WITNESS/TA	DATE

NOTE: INSERT DIVIDER UNDER COPY SHEET BEFORE WRITING

EXP. NUMBER	EXPERIMENT/SUBJECT		DATE	
NAME		LAB PARTNER	LOCKER/DESK NO.	COURSE & SECTION NO.

SIGNATURE	DATE	WITNESS/TA	DATE

EXP. NUMBER	EXPERIMENT/SUBJECT		DATE	
NAME		LAB PARTNER	LOCKER/DESK NO.	COURSE & SECTION NO.

EXP. NUMBER	EXPERIMENT/SUBJECT		DATE	
SIGNATURE		DATE	WITNESS/TA	DATE

EXP. NUMBER	EXPERIMENT/SUBJECT		DATE	
NAME		LAB PARTNER	LOCKER/DESK NO.	COURSE & SECTION NO.

SIGNATURE	DATE	WITNESS/TA	DATE

THE HAYDEN-McNEIL STUDENT LAB NOTEBOOK

EXP. NUMBER	EXPERIMENT/SUBJECT		DATE	
NAME		LAB PARTNER	LOCKER/DESK NO.	COURSE & SECTION NO.

SIGNATURE		DATE	WITNESS/TA		DATE

NOTE: INSERT DIVIDER UNDER COPY SHEET BEFORE WRITING

EXP. NUMBER	EXPERIMENT/SUBJECT		DATE	
NAME		LAB PARTNER	LOCKER/DESK NO.	COURSE & SECTION NO.

SIGNATURE		DATE	WITNESS/TA		DATE

NOTE: INSERT DIVIDER UNDER COPY SHEET BEFORE WRITING

EXP. NUMBER	EXPERIMENT/SUBJECT		DATE	
NAME		LAB PARTNER	LOCKER/DESK NO.	COURSE & SECTION NO.

COPY

SIGNATURE		DATE	WITNESS/TA		DATE

NOTE: INSERT DIVIDER UNDER COPY SHEET BEFORE WRITING

EXP. NUMBER	EXPERIMENT/SUBJECT		DATE	
NAME		LAB PARTNER	LOCKER/DESK NO.	COURSE & SECTION NO.

SIGNATURE		DATE	WITNESS/TA		DATE

EXP. NUMBER	EXPERIMENT/SUBJECT		DATE	
NAME		LAB PARTNER	LOCKER/DESK NO.	COURSE & SECTION NO.

SIGNATURE		DATE	WITNESS/TA		DATE

NOTE: INSERT DIVIDER UNDER COPY SHEET BEFORE WRITING

EXP. NUMBER	EXPERIMENT/SUBJECT		DATE	
NAME		LAB PARTNER	LOCKER/DESK NO.	COURSE & SECTION NO.

SIGNATURE		DATE	WITNESS/TA		DATE

EXP. NUMBER	EXPERIMENT/SUBJECT		DATE	
NAME		LAB PARTNER	LOCKER/DESK NO.	COURSE & SECTION NO.

SIGNATURE	DATE	WITNESS/TA		DATE

EXP. NUMBER	EXPERIMENT/SUBJECT		DATE	
NAME		LAB PARTNER	LOCKER/DESK NO.	COURSE & SECTION NO.

SIGNATURE	DATE	WITNESS/TA		DATE

NOTE: INSERT DIVIDER UNDER COPY SHEET BEFORE WRITING

EXP. NUMBER	EXPERIMENT/SUBJECT		DATE	
NAME		LAB PARTNER	LOCKER/DESK NO.	COURSE & SECTION NO.

EXP. NUMBER	EXPERIMENT/SUBJECT		DATE	
SIGNATURE		DATE	WITNESS/TA	DATE

EXP. NUMBER	EXPERIMENT/SUBJECT		DATE	
NAME		LAB PARTNER	LOCKER/DESK NO.	COURSE & SECTION NO.

EXP. NUMBER	EXPERIMENT/SUBJECT		DATE	

SIGNATURE	DATE	WITNESS/TA	DATE

EXP. NUMBER	EXPERIMENT/SUBJECT		DATE	
NAME		LAB PARTNER	LOCKER/DESK NO.	COURSE & SECTION NO.

			DATE	
EXP. NUMBER	EXPERIMENT/SUBJECT			
SIGNATURE		DATE	WITNESS/TA	DATE

NOTE: INSERT DIVIDER UNDER COPY SHEET BEFORE WRITING

EXP. NUMBER	EXPERIMENT/SUBJECT		DATE	
NAME		LAB PARTNER	LOCKER/DESK NO.	COURSE & SECTION NO.

SIGNATURE		DATE	WITNESS/TA		DATE

NOTE: INSERT DIVIDER UNDER COPY SHEET BEFORE WRITING

EXP. NUMBER	EXPERIMENT/SUBJECT		DATE	
NAME		LAB PARTNER	LOCKER/DESK NO.	COURSE & SECTION NO.

SIGNATURE	DATE	WITNESS/TA	DATE

NOTE: INSERT DIVIDER UNDER COPY SHEET BEFORE WRITING

EXP. NUMBER	EXPERIMENT/SUBJECT		DATE	
NAME		LAB PARTNER	LOCKER/DESK NO.	COURSE & SECTION NO.

SIGNATURE	DATE	WITNESS/TA	DATE

THE HAYDEN-McNEIL STUDENT LAB NOTEBOOK

NOTE: INSERT DIVIDER UNDER COPY SHEET BEFORE WRITING

EXP. NUMBER	EXPERIMENT/SUBJECT		DATE	
NAME		LAB PARTNER	LOCKER/DESK NO.	COURSE & SECTION NO.

36

SIGNATURE	DATE	WITNESS/TA	DATE

NOTE: INSERT DIVIDER UNDER COPY SHEET BEFORE WRITING

EXP. NUMBER	EXPERIMENT/SUBJECT		DATE	
NAME		LAB PARTNER	LOCKER/DESK NO.	COURSE & SECTION NO.

SIGNATURE	DATE	WITNESS/TA	DATE

THE HAYDEN-McNEIL STUDENT LAB NOTEBOOK

NOTE: INSERT DIVIDER UNDER COPY SHEET BEFORE WRITING

EXP. NUMBER	EXPERIMENT/SUBJECT		DATE	
NAME		LAB PARTNER	LOCKER/DESK NO.	COURSE & SECTION NO.

SIGNATURE	DATE	WITNESS/TA	DATE

NOTE: INSERT DIVIDER UNDER COPY SHEET BEFORE WRITING

EXP. NUMBER	EXPERIMENT/SUBJECT		DATE	
NAME		LAB PARTNER	LOCKER/DESK NO.	COURSE & SECTION NO.

SIGNATURE	DATE	WITNESS/TA	DATE

EXP. NUMBER	EXPERIMENT/SUBJECT		DATE	
NAME		LAB PARTNER	LOCKER/DESK NO.	COURSE & SECTION NO.

SIGNATURE	DATE	WITNESS/TA	DATE

NOTE: INSERT DIVIDER UNDER COPY SHEET BEFORE WRITING

EXP. NUMBER	EXPERIMENT/SUBJECT		DATE	
NAME		LAB PARTNER	LOCKER/DESK NO.	COURSE & SECTION NO.

SIGNATURE	DATE	WITNESS/TA		DATE

EXP. NUMBER	EXPERIMENT/SUBJECT		DATE	
NAME		LAB PARTNER	LOCKER/DESK NO.	COURSE & SECTION NO.

SIGNATURE	DATE	WITNESS/TA	DATE

NOTE: INSERT DIVIDER UNDER COPY SHEET BEFORE WRITING

EXP. NUMBER	EXPERIMENT/SUBJECT		DATE	
NAME		LAB PARTNER	LOCKER/DESK NO.	COURSE & SECTION NO.

SIGNATURE	DATE	WITNESS/TA	DATE

THE HAYDEN-McNEIL STUDENT LAB NOTEBOOK

NOTE: INSERT DIVIDER UNDER COPY SHEET BEFORE WRITING

EXP. NUMBER	EXPERIMENT/SUBJECT		DATE	
NAME		LAB PARTNER	LOCKER/DESK NO.	COURSE & SECTION NO.

SIGNATURE	DATE	WITNESS/TA	DATE

NOTE: INSERT DIVIDER UNDER COPY SHEET BEFORE WRITING

EXP. NUMBER	EXPERIMENT/SUBJECT		DATE	
NAME		LAB PARTNER	LOCKER/DESK NO.	COURSE & SECTION NO.

SIGNATURE	DATE	WITNESS/TA	DATE

EXP. NUMBER	EXPERIMENT/SUBJECT		DATE	
NAME		LAB PARTNER	LOCKER/DESK NO.	COURSE & SECTION NO.

EXP. NUMBER	EXPERIMENT/SUBJECT		DATE	

SIGNATURE		DATE	WITNESS/TA		DATE

EXP. NUMBER	EXPERIMENT/SUBJECT		DATE	
NAME		LAB PARTNER	LOCKER/DESK NO.	COURSE & SECTION NO.

SIGNATURE		DATE	WITNESS/TA		DATE

EXP. NUMBER	EXPERIMENT/SUBJECT		DATE	
NAME		LAB PARTNER	LOCKER/DESK NO.	COURSE & SECTION NO.

SIGNATURE		DATE	WITNESS/TA		DATE

NOTE: INSERT DIVIDER UNDER COPY SHEET BEFORE WRITING

EXP. NUMBER	EXPERIMENT/SUBJECT		DATE	
NAME		LAB PARTNER	LOCKER/DESK NO.	COURSE & SECTION NO.

SIGNATURE		DATE	WITNESS/TA		DATE

EXP. NUMBER	EXPERIMENT/SUBJECT		DATE	
NAME		LAB PARTNER	LOCKER/DESK NO.	COURSE & SECTION NO.

43

SIGNATURE	DATE	WITNESS/TA	DATE

THE HAYDEN-McNEIL STUDENT LAB NOTEBOOK

NOTE: INSERT DIVIDER UNDER COPY SHEET BEFORE WRITING

EXP. NUMBER	EXPERIMENT/SUBJECT		DATE	
NAME		LAB PARTNER	LOCKER/DESK NO.	COURSE & SECTION NO.

SIGNATURE	DATE	WITNESS/TA	DATE

NOTE: INSERT DIVIDER UNDER COPY SHEET BEFORE WRITING

EXP. NUMBER	EXPERIMENT/SUBJECT		DATE	
NAME		LAB PARTNER	LOCKER/DESK NO.	COURSE & SECTION NO.

SIGNATURE	DATE	WITNESS/TA	DATE

NOTE: INSERT DIVIDER UNDER COPY SHEET BEFORE WRITING

EXP. NUMBER	EXPERIMENT/SUBJECT		DATE	
NAME		LAB PARTNER	LOCKER/DESK NO.	COURSE & SECTION NO.

SIGNATURE	DATE	WITNESS/TA		DATE

EXP. NUMBER	EXPERIMENT/SUBJECT		DATE	
NAME		LAB PARTNER	LOCKER/DESK NO.	COURSE & SECTION NO.

SIGNATURE	DATE	WITNESS/TA	DATE

NOTE: INSERT DIVIDER UNDER COPY SHEET BEFORE WRITING

EXP. NUMBER	EXPERIMENT/SUBJECT		DATE	
NAME		LAB PARTNER	LOCKER/DESK NO.	COURSE & SECTION NO.

SIGNATURE	DATE	WITNESS/TA	DATE

NOTE: INSERT DIVIDER UNDER COPY SHEET BEFORE WRITING

EXP. NUMBER	EXPERIMENT/SUBJECT		DATE	
NAME		LAB PARTNER	LOCKER/DESK NO.	COURSE & SECTION NO.

SIGNATURE	DATE	WITNESS/TA	DATE

NOTE: INSERT DIVIDER UNDER COPY SHEET BEFORE WRITING

EXP. NUMBER	EXPERIMENT/SUBJECT		DATE	
NAME		LAB PARTNER	LOCKER/DESK NO.	COURSE & SECTION NO.

SIGNATURE	DATE	WITNESS/TA		DATE

NOTE: INSERT DIVIDER UNDER COPY SHEET BEFORE WRITING

EXP. NUMBER	EXPERIMENT/SUBJECT		DATE	
NAME		LAB PARTNER	LOCKER/DESK NO.	COURSE & SECTION NO.

SIGNATURE	DATE	WITNESS/TA	DATE

NOTE: INSERT DIVIDER UNDER COPY SHEET BEFORE WRITING

EXP. NUMBER	EXPERIMENT/SUBJECT		DATE	
NAME		LAB PARTNER	LOCKER/DESK NO.	COURSE & SECTION NO.

SIGNATURE	DATE	WITNESS/TA	DATE

EXP. NUMBER	EXPERIMENT/SUBJECT		DATE	
NAME		LAB PARTNER	LOCKER/DESK NO.	COURSE & SECTION NO.

48

SIGNATURE	DATE	WITNESS/TA		DATE

NOTE: INSERT DIVIDER UNDER COPY SHEET BEFORE WRITING

EXP. NUMBER	EXPERIMENT/SUBJECT		DATE	
NAME		LAB PARTNER	LOCKER/DESK NO.	COURSE & SECTION NO.

SIGNATURE	DATE	WITNESS/TA		DATE

THE HAYDEN-McNEIL STUDENT LAB NOTEBOOK

NOTE: INSERT DIVIDER UNDER COPY SHEET BEFORE WRITING

EXP. NUMBER	EXPERIMENT/SUBJECT		DATE	
NAME		LAB PARTNER	LOCKER/DESK NO.	COURSE & SECTION NO.

SIGNATURE	DATE	WITNESS/TA	DATE

NOTE: INSERT DIVIDER UNDER COPY SHEET BEFORE WRITING

| EXP. NUMBER | EXPERIMENT/SUBJECT | | DATE | |
| NAME | | LAB PARTNER | LOCKER/DESK NO. | COURSE & SECTION NO. |

| SIGNATURE | | DATE | WITNESS/TA | | DATE |

EXP. NUMBER	EXPERIMENT/SUBJECT		DATE	
NAME		LAB PARTNER	LOCKER/DESK NO.	COURSE & SECTION NO.

SIGNATURE	DATE	WITNESS/TA	DATE

EXP. NUMBER	EXPERIMENT/SUBJECT		DATE	
NAME		LAB PARTNER	LOCKER/DESK NO.	COURSE & SECTION NO.

SIGNATURE	DATE	WITNESS/TA	DATE

THE HAYDEN-McNEIL STUDENT LAB NOTEBOOK

EXP. NUMBER	EXPERIMENT/SUBJECT		DATE	
NAME		LAB PARTNER	LOCKER/DESK NO.	COURSE & SECTION NO.

SIGNATURE		DATE	WITNESS/TA		DATE

NOTE: INSERT DIVIDER UNDER COPY SHEET BEFORE WRITING

EXP. NUMBER	EXPERIMENT/SUBJECT		DATE	
NAME		LAB PARTNER	LOCKER/DESK NO.	COURSE & SECTION NO.

SIGNATURE		DATE	WITNESS/TA		DATE

NOTE: INSERT DIVIDER UNDER COPY SHEET BEFORE WRITING

EXP. NUMBER	EXPERIMENT/SUBJECT		DATE	
NAME		LAB PARTNER	LOCKER/DESK NO.	COURSE & SECTION NO.

SIGNATURE		DATE	WITNESS/TA	DATE

NOTE: INSERT DIVIDER UNDER COPY SHEET BEFORE WRITING

EXP. NUMBER	EXPERIMENT/SUBJECT		DATE	
NAME		LAB PARTNER	LOCKER/DESK NO.	COURSE & SECTION NO.

SIGNATURE	DATE	WITNESS/TA	DATE

THE HAYDEN-McNEIL STUDENT LAB NOTEBOOK

EXP. NUMBER	EXPERIMENT/SUBJECT		DATE	
NAME		LAB PARTNER	LOCKER/DESK NO.	COURSE & SECTION NO.

SIGNATURE	DATE	WITNESS/TA	DATE

NOTE: INSERT DIVIDER UNDER COPY SHEET BEFORE WRITING

EXP. NUMBER	EXPERIMENT/SUBJECT		DATE	
NAME		LAB PARTNER	LOCKER/DESK NO.	COURSE & SECTION NO.

SIGNATURE	DATE	WITNESS/TA	DATE

EXP. NUMBER	EXPERIMENT/SUBJECT		DATE	
NAME		LAB PARTNER	LOCKER/DESK NO.	COURSE & SECTION NO.

SIGNATURE	DATE	WITNESS/TA	DATE

NOTE: INSERT DIVIDER UNDER COPY SHEET BEFORE WRITING

EXP. NUMBER	EXPERIMENT/SUBJECT		DATE	
NAME		LAB PARTNER	LOCKER/DESK NO.	COURSE & SECTION NO.

SIGNATURE	DATE	WITNESS/TA	DATE

EXP. NUMBER	EXPERIMENT/SUBJECT		DATE	
NAME		LAB PARTNER	LOCKER/DESK NO.	COURSE & SECTION NO.

SIGNATURE		DATE	WITNESS/TA		DATE

THE HAYDEN-McNEIL STUDENT LAB NOTEBOOK

NOTE: INSERT DIVIDER UNDER COPY SHEET BEFORE WRITING

EXP. NUMBER	EXPERIMENT/SUBJECT		DATE	
NAME		LAB PARTNER	LOCKER/DESK NO.	COURSE & SECTION NO.

SIGNATURE	DATE	WITNESS/TA		DATE

NOTE: INSERT DIVIDER UNDER COPY SHEET BEFORE WRITING

56

EXP. NUMBER	EXPERIMENT/SUBJECT		DATE	
NAME		LAB PARTNER	LOCKER/DESK NO.	COURSE & SECTION NO.

SIGNATURE	DATE	WITNESS/TA	DATE

THE HAYDEN-McNEIL STUDENT LAB NOTEBOOK

NOTE: INSERT DIVIDER UNDER COPY SHEET BEFORE WRITING

EXP. NUMBER	EXPERIMENT/SUBJECT		DATE	
NAME		LAB PARTNER	LOCKER/DESK NO.	COURSE & SECTION NO.

SIGNATURE	DATE	WITNESS/TA	DATE

NOTE: INSERT DIVIDER UNDER COPY SHEET BEFORE WRITING

EXP. NUMBER	EXPERIMENT/SUBJECT		DATE	
NAME		LAB PARTNER	LOCKER/DESK NO.	COURSE & SECTION NO.

SIGNATURE		DATE	WITNESS/TA		DATE

EXP. NUMBER	EXPERIMENT/SUBJECT		DATE	
NAME		LAB PARTNER	LOCKER/DESK NO.	COURSE & SECTION NO.

EXP. NUMBER	EXPERIMENT/SUBJECT		DATE

SIGNATURE	DATE	WITNESS/TA	DATE

NOTE: INSERT DIVIDER UNDER COPY SHEET BEFORE WRITING

EXP. NUMBER	EXPERIMENT/SUBJECT		DATE	
NAME		LAB PARTNER	LOCKER/DESK NO.	COURSE & SECTION NO.

SIGNATURE		DATE	WITNESS/TA		DATE

NOTE: INSERT DIVIDER UNDER COPY SHEET BEFORE WRITING

EXP. NUMBER	EXPERIMENT/SUBJECT		DATE	
NAME		LAB PARTNER	LOCKER/DESK NO.	COURSE & SECTION NO.

SIGNATURE		DATE	WITNESS/TA	DATE

NOTE: INSERT DIVIDER UNDER COPY SHEET BEFORE WRITING

EXP. NUMBER	EXPERIMENT/SUBJECT		DATE	
NAME		LAB PARTNER	LOCKER/DESK NO.	COURSE & SECTION NO.

SIGNATURE	DATE	WITNESS/TA	DATE

EXP. NUMBER	EXPERIMENT/SUBJECT		DATE	
NAME		LAB PARTNER	LOCKER/DESK NO.	COURSE & SECTION NO.

SIGNATURE	DATE	WITNESS/TA	DATE

NOTE: INSERT DIVIDER UNDER COPY SHEET BEFORE WRITING

EXP. NUMBER	EXPERIMENT/SUBJECT		DATE	
NAME		LAB PARTNER	LOCKER/DESK NO.	COURSE & SECTION NO.

SIGNATURE	DATE	WITNESS/TA	DATE

NOTE: INSERT DIVIDER UNDER COPY SHEET BEFORE WRITING

EXP. NUMBER	EXPERIMENT/SUBJECT		DATE	
NAME		LAB PARTNER	LOCKER/DESK NO.	COURSE & SECTION NO.

SIGNATURE	DATE	WITNESS/TA		DATE

NOTE: INSERT DIVIDER UNDER COPY SHEET BEFORE WRITING

EXP. NUMBER	EXPERIMENT/SUBJECT		DATE	
NAME		LAB PARTNER	LOCKER/DESK NO.	COURSE & SECTION NO.

SIGNATURE	DATE	WITNESS/TA		DATE

EXP. NUMBER	EXPERIMENT/SUBJECT		DATE	
NAME		LAB PARTNER	LOCKER/DESK NO.	COURSE & SECTION NO.

SIGNATURE	DATE	WITNESS/TA	DATE

NOTE: INSERT DIVIDER UNDER COPY SHEET BEFORE WRITING

EXP. NUMBER	EXPERIMENT/SUBJECT		DATE	
NAME		LAB PARTNER	LOCKER/DESK NO.	COURSE & SECTION NO.

SIGNATURE		DATE	WITNESS/TA		DATE

EXP. NUMBER	EXPERIMENT/SUBJECT		DATE	
NAME		LAB PARTNER	LOCKER/DESK NO.	COURSE & SECTION NO.

SIGNATURE		DATE	WITNESS/TA		DATE

NOTE: INSERT DIVIDER UNDER COPY SHEET BEFORE WRITING

EXP. NUMBER	EXPERIMENT/SUBJECT		DATE	
NAME		LAB PARTNER	LOCKER/DESK NO.	COURSE & SECTION NO.

SIGNATURE		DATE	WITNESS/TA		DATE

THE HAYDEN-McNEIL STUDENT LAB NOTEBOOK

EXP. NUMBER	EXPERIMENT/SUBJECT		DATE	
NAME		LAB PARTNER	LOCKER/DESK NO.	COURSE & SECTION NO.

SIGNATURE	DATE	WITNESS/TA	DATE

NOTE: INSERT DIVIDER UNDER COPY SHEET BEFORE WRITING

EXP. NUMBER	EXPERIMENT/SUBJECT		DATE	
NAME		LAB PARTNER	LOCKER/DESK NO.	COURSE & SECTION NO.

SIGNATURE	DATE	WITNESS/TA	DATE

NOTE: INSERT DIVIDER UNDER COPY SHEET BEFORE WRITING

EXP. NUMBER	EXPERIMENT/SUBJECT		DATE	
NAME		LAB PARTNER	LOCKER/DESK NO.	COURSE & SECTION NO.

SIGNATURE	DATE	WITNESS/TA	DATE

EXP. NUMBER	EXPERIMENT/SUBJECT		DATE	
NAME		LAB PARTNER	LOCKER/DESK NO.	COURSE & SECTION NO.

EXP. NUMBER	EXPERIMENT/SUBJECT		DATE	
SIGNATURE		DATE	WITNESS/TA	DATE

NOTE: INSERT DIVIDER UNDER COPY SHEET BEFORE WRITING

EXP. NUMBER	EXPERIMENT/SUBJECT		DATE	
NAME		LAB PARTNER	LOCKER/DESK NO.	COURSE & SECTION NO.

SIGNATURE	DATE	WITNESS/TA	DATE

EXP. NUMBER	EXPERIMENT/SUBJECT		DATE	
NAME		LAB PARTNER	LOCKER/DESK NO.	COURSE & SECTION NO.

SIGNATURE	DATE	WITNESS/TA	DATE

NOTE: INSERT DIVIDER UNDER COPY SHEET BEFORE WRITING

EXP. NUMBER	EXPERIMENT/SUBJECT		DATE	
NAME		LAB PARTNER	LOCKER/DESK NO.	COURSE & SECTION NO.

SIGNATURE	DATE	WITNESS/TA	DATE

THE HAYDEN-McNEIL STUDENT LAB NOTEBOOK

NOTE: INSERT DIVIDER UNDER COPY SHEET BEFORE WRITING

EXP. NUMBER	EXPERIMENT/SUBJECT		DATE	
NAME		LAB PARTNER	LOCKER/DESK NO.	COURSE & SECTION NO.

SIGNATURE	DATE	WITNESS/TA	DATE

NOTE: INSERT DIVIDER UNDER COPY SHEET BEFORE WRITING

EXP. NUMBER	EXPERIMENT/SUBJECT		DATE	
NAME		LAB PARTNER	LOCKER/DESK NO.	COURSE & SECTION NO.

SIGNATURE	DATE	WITNESS/TA	DATE

NOTE: INSERT DIVIDER UNDER COPY SHEET BEFORE WRITING

EXP. NUMBER	EXPERIMENT/SUBJECT		DATE	
NAME		LAB PARTNER	LOCKER/DESK NO.	COURSE & SECTION NO.

SIGNATURE		DATE	WITNESS/TA		DATE

EXP. NUMBER	EXPERIMENT/SUBJECT		DATE	
NAME		LAB PARTNER	LOCKER/DESK NO.	COURSE & SECTION NO.

SIGNATURE		DATE	WITNESS/TA		DATE

THE HAYDEN-McNEIL STUDENT LAB NOTEBOOK

NOTE: INSERT DIVIDER UNDER COPY SHEET BEFORE WRITING

EXP. NUMBER	EXPERIMENT/SUBJECT		DATE	
NAME		LAB PARTNER	LOCKER/DESK NO.	COURSE & SECTION NO.

SIGNATURE	DATE	WITNESS/TA		DATE

NOTE: INSERT DIVIDER UNDER COPY SHEET BEFORE WRITING

EXP. NUMBER	EXPERIMENT/SUBJECT		DATE	
NAME		LAB PARTNER	LOCKER/DESK NO.	COURSE & SECTION NO.

SIGNATURE	DATE	WITNESS/TA	DATE

EXP. NUMBER	EXPERIMENT/SUBJECT		DATE	
NAME		LAB PARTNER	LOCKER/DESK NO.	COURSE & SECTION NO.

SIGNATURE		DATE	WITNESS/TA		DATE

NOTE: INSERT DIVIDER UNDER COPY SHEET BEFORE WRITING

EXP. NUMBER	EXPERIMENT/SUBJECT		DATE	
NAME		LAB PARTNER	LOCKER/DESK NO.	COURSE & SECTION NO.

SIGNATURE		DATE	WITNESS/TA		DATE

THE HAYDEN-McNEIL STUDENT LAB NOTEBOOK

NOTE: INSERT DIVIDER UNDER COPY SHEET BEFORE WRITING

EXP. NUMBER	EXPERIMENT/SUBJECT		DATE	
NAME		LAB PARTNER	LOCKER/DESK NO.	COURSE & SECTION NO.

71

COPY

SIGNATURE	DATE	WITNESS/TA	DATE

THE HAYDEN-McNEIL STUDENT LAB NOTEBOOK

NOTE: INSERT DIVIDER UNDER COPY SHEET BEFORE WRITING

EXP. NUMBER	EXPERIMENT/SUBJECT		DATE	
NAME		LAB PARTNER	LOCKER/DESK NO.	COURSE & SECTION NO.

SIGNATURE	DATE	WITNESS/TA	DATE

EXP. NUMBER	EXPERIMENT/SUBJECT		DATE	
NAME		LAB PARTNER	LOCKER/DESK NO.	COURSE & SECTION NO.

SIGNATURE	DATE	WITNESS/TA	DATE

NOTE: INSERT DIVIDER UNDER COPY SHEET BEFORE WRITING

EXP. NUMBER	EXPERIMENT/SUBJECT		DATE	
NAME		LAB PARTNER	LOCKER/DESK NO.	COURSE & SECTION NO.

SIGNATURE	DATE	WITNESS/TA	DATE

EXP. NUMBER	EXPERIMENT/SUBJECT		DATE	
NAME		LAB PARTNER	LOCKER/DESK NO.	COURSE & SECTION NO.

SIGNATURE	DATE	WITNESS/TA	DATE

NOTE: INSERT DIVIDER UNDER COPY SHEET BEFORE WRITING

EXP. NUMBER	EXPERIMENT/SUBJECT		DATE	
NAME		LAB PARTNER	LOCKER/DESK NO.	COURSE & SECTION NO.

SIGNATURE	DATE	WITNESS/TA	DATE

NOTE: INSERT DIVIDER UNDER COPY SHEET BEFORE WRITING

EXP. NUMBER	EXPERIMENT/SUBJECT		DATE	
NAME		LAB PARTNER	LOCKER/DESK NO.	COURSE & SECTION NO.

SIGNATURE	DATE	WITNESS/TA	DATE

THE HAYDEN-McNEIL STUDENT LAB NOTEBOOK

75

EXP. NUMBER	EXPERIMENT/SUBJECT		DATE	
NAME		LAB PARTNER	LOCKER/DESK NO.	COURSE & SECTION NO.

SIGNATURE	DATE	WITNESS/TA	DATE

THE HAYDEN-McNEIL STUDENT LAB NOTEBOOK

NOTE: INSERT DIVIDER UNDER COPY SHEET BEFORE WRITING

EXP. NUMBER	EXPERIMENT/SUBJECT		DATE	
NAME		LAB PARTNER	LOCKER/DESK NO.	COURSE & SECTION NO.

SIGNATURE	DATE	WITNESS/TA	DATE

EXP. NUMBER	EXPERIMENT/SUBJECT		DATE	
NAME		LAB PARTNER	LOCKER/DESK NO.	COURSE & SECTION NO.

SIGNATURE	DATE	WITNESS/TA	DATE

NOTE: INSERT DIVIDER UNDER COPY SHEET BEFORE WRITING

EXP. NUMBER	EXPERIMENT/SUBJECT		DATE	
NAME		LAB PARTNER	LOCKER/DESK NO.	COURSE & SECTION NO.

SIGNATURE	DATE	WITNESS/TA	DATE

NOTE: INSERT DIVIDER UNDER COPY SHEET BEFORE WRITING

EXP. NUMBER	EXPERIMENT/SUBJECT		DATE	
NAME		LAB PARTNER	LOCKER/DESK NO.	COURSE & SECTION NO.

SIGNATURE	DATE	WITNESS/TA		DATE

EXP. NUMBER	EXPERIMENT/SUBJECT		DATE	
NAME		LAB PARTNER	LOCKER/DESK NO.	COURSE & SECTION NO.

SIGNATURE	DATE	WITNESS/TA	DATE

NOTE: INSERT DIVIDER UNDER COPY SHEET BEFORE WRITING

EXP. NUMBER	EXPERIMENT/SUBJECT		DATE	
NAME		LAB PARTNER	LOCKER/DESK NO.	COURSE & SECTION NO.

SIGNATURE	DATE	WITNESS/TA	DATE

NOTE: INSERT DIVIDER UNDER COPY SHEET BEFORE WRITING

EXP. NUMBER	EXPERIMENT/SUBJECT		DATE	
NAME		LAB PARTNER	LOCKER/DESK NO.	COURSE & SECTION NO.

SIGNATURE	DATE	WITNESS/TA	DATE

NOTE: INSERT DIVIDER UNDER COPY SHEET BEFORE WRITING

EXP. NUMBER	EXPERIMENT/SUBJECT		DATE	
NAME		LAB PARTNER	LOCKER/DESK NO.	COURSE & SECTION NO.

SIGNATURE		DATE	WITNESS/TA		DATE

NOTE: INSERT DIVIDER UNDER COPY SHEET BEFORE WRITING

EXP. NUMBER	EXPERIMENT/SUBJECT		DATE	
NAME		LAB PARTNER	LOCKER/DESK NO.	COURSE & SECTION NO.

SIGNATURE	DATE	WITNESS/TA	DATE

EXP. NUMBER	EXPERIMENT/SUBJECT		DATE	
NAME		LAB PARTNER	LOCKER/DESK NO.	COURSE & SECTION NO.

EXP. NUMBER	EXPERIMENT/SUBJECT		DATE	
SIGNATURE		DATE	WITNESS/TA	DATE

EXP. NUMBER	EXPERIMENT/SUBJECT		DATE	
NAME		LAB PARTNER	LOCKER/DESK NO.	COURSE & SECTION NO.

SIGNATURE	DATE	WITNESS/TA	DATE

NOTE: INSERT DIVIDER UNDER COPY SHEET BEFORE WRITING

EXP. NUMBER	EXPERIMENT/SUBJECT		DATE	
NAME		LAB PARTNER	LOCKER/DESK NO.	COURSE & SECTION NO.

SIGNATURE	DATE	WITNESS/TA	DATE

EXP. NUMBER	EXPERIMENT/SUBJECT		DATE	
NAME		LAB PARTNER	LOCKER/DESK NO.	COURSE & SECTION NO.

SIGNATURE	DATE	WITNESS/TA	DATE

NOTE: INSERT DIVIDER UNDER COPY SHEET BEFORE WRITING

EXP. NUMBER	EXPERIMENT/SUBJECT		DATE	
NAME		LAB PARTNER	LOCKER/DESK NO.	COURSE & SECTION NO.

SIGNATURE	DATE	WITNESS/TA	DATE

EXP. NUMBER	EXPERIMENT/SUBJECT		DATE	
NAME		LAB PARTNER	LOCKER/DESK NO.	COURSE & SECTION NO.

SIGNATURE	DATE	WITNESS/TA	DATE

NOTE: INSERT DIVIDER UNDER COPY SHEET BEFORE WRITING

EXP. NUMBER	EXPERIMENT/SUBJECT		DATE	
NAME		LAB PARTNER	LOCKER/DESK NO.	COURSE & SECTION NO.

SIGNATURE	DATE	WITNESS/TA		DATE

THE HAYDEN-McNEIL STUDENT LAB NOTEBOOK

NOTE: INSERT DIVIDER UNDER COPY SHEET BEFORE WRITING

EXP. NUMBER	EXPERIMENT/SUBJECT		DATE	
NAME		LAB PARTNER	LOCKER/DESK NO.	COURSE & SECTION NO.

SIGNATURE	DATE	WITNESS/TA		DATE

NOTE: INSERT DIVIDER UNDER COPY SHEET BEFORE WRITING

EXP. NUMBER	EXPERIMENT/SUBJECT		DATE	
NAME		LAB PARTNER	LOCKER/DESK NO.	COURSE & SECTION NO.

SIGNATURE	DATE	WITNESS/TA	DATE

EXP. NUMBER	EXPERIMENT/SUBJECT		DATE	
NAME		LAB PARTNER	LOCKER/DESK NO.	COURSE & SECTION NO.

84

SIGNATURE	DATE	WITNESS/TA	DATE

EXP. NUMBER	EXPERIMENT/SUBJECT		DATE	
NAME		LAB PARTNER	LOCKER/DESK NO.	COURSE & SECTION NO.

SIGNATURE	DATE	WITNESS/TA		DATE

THE HAYDEN-McNEIL STUDENT LAB NOTEBOOK

NOTE: INSERT DIVIDER UNDER COPY SHEET BEFORE WRITING

EXP. NUMBER	EXPERIMENT/SUBJECT		DATE	
NAME		LAB PARTNER	LOCKER/DESK NO.	COURSE & SECTION NO.

SIGNATURE	DATE	WITNESS/TA	DATE

NOTE: INSERT DIVIDER UNDER COPY SHEET BEFORE WRITING

EXP. NUMBER	EXPERIMENT/SUBJECT		DATE	
NAME		LAB PARTNER	LOCKER/DESK NO.	COURSE & SECTION NO.

SIGNATURE	DATE	WITNESS/TA	DATE

THE HAYDEN-McNEIL STUDENT LAB NOTEBOOK

NOTE: INSERT DIVIDER UNDER COPY SHEET BEFORE WRITING

EXP. NUMBER	EXPERIMENT/SUBJECT		DATE	
NAME		LAB PARTNER	LOCKER/DESK NO.	COURSE & SECTION NO.

SIGNATURE	DATE	WITNESS/TA		DATE

NOTE: INSERT DIVIDER UNDER COPY SHEET BEFORE WRITING

EXP. NUMBER	EXPERIMENT/SUBJECT		DATE	
NAME		LAB PARTNER	LOCKER/DESK NO.	COURSE & SECTION NO.

SIGNATURE	DATE	WITNESS/TA	DATE

EXP. NUMBER	EXPERIMENT/SUBJECT		DATE	
NAME	LAB PARTNER		LOCKER/DESK NO.	COURSE & SECTION NO.

SIGNATURE	DATE	WITNESS/TA		DATE

NOTE: INSERT DIVIDER UNDER COPY SHEET BEFORE WRITING

EXP. NUMBER	EXPERIMENT/SUBJECT		DATE	
NAME		LAB PARTNER	LOCKER/DESK NO.	COURSE & SECTION NO.

SIGNATURE	DATE	WITNESS/TA	DATE

NOTE: INSERT DIVIDER UNDER COPY SHEET BEFORE WRITING

88

EXP. NUMBER	EXPERIMENT/SUBJECT		DATE	
NAME		LAB PARTNER	LOCKER/DESK NO.	COURSE & SECTION NO.

SIGNATURE	DATE	WITNESS/TA	DATE

THE HAYDEN-McNEIL STUDENT LAB NOTEBOOK

NOTE: INSERT DIVIDER UNDER COPY SHEET BEFORE WRITING

89

EXP. NUMBER	EXPERIMENT/SUBJECT		DATE
NAME	LAB PARTNER	LOCKER/DESK NO.	COURSE & SECTION NO.

SIGNATURE	DATE	WITNESS/TA	DATE

NOTE: INSERT DIVIDER UNDER COPY SHEET BEFORE WRITING

EXP. NUMBER	EXPERIMENT/SUBJECT		DATE	
NAME		LAB PARTNER	LOCKER/DESK NO.	COURSE & SECTION NO.

EXP. NUMBER	EXPERIMENT/SUBJECT		DATE	
SIGNATURE		DATE	WITNESS/TA	DATE

NOTE: INSERT DIVIDER UNDER COPY SHEET BEFORE WRITING

EXP. NUMBER	EXPERIMENT/SUBJECT		DATE	
NAME		LAB PARTNER	LOCKER/DESK NO.	COURSE & SECTION NO.

SIGNATURE	DATE	WITNESS/TA	DATE

EXP. NUMBER	EXPERIMENT/SUBJECT		DATE	
NAME		LAB PARTNER	LOCKER/DESK NO.	COURSE & SECTION NO.

SIGNATURE	DATE	WITNESS/TA	DATE

NOTE: INSERT DIVIDER UNDER COPY SHEET BEFORE WRITING

EXP. NUMBER	EXPERIMENT/SUBJECT		DATE	
NAME		LAB PARTNER	LOCKER/DESK NO.	COURSE & SECTION NO.

SIGNATURE		DATE	WITNESS/TA		DATE

EXP. NUMBER	EXPERIMENT/SUBJECT		DATE	
NAME		LAB PARTNER	LOCKER/DESK NO.	COURSE & SECTION NO.

SIGNATURE	DATE	WITNESS/TA	DATE

NOTE: INSERT DIVIDER UNDER COPY SHEET BEFORE WRITING

EXP. NUMBER	EXPERIMENT/SUBJECT		DATE	
NAME		LAB PARTNER	LOCKER/DESK NO.	COURSE & SECTION NO.

SIGNATURE	DATE	WITNESS/TA	DATE

EXP. NUMBER	EXPERIMENT/SUBJECT		DATE	
NAME		LAB PARTNER	LOCKER/DESK NO.	COURSE & SECTION NO.

SIGNATURE	DATE	WITNESS/TA	DATE

NOTE: INSERT DIVIDER UNDER COPY SHEET BEFORE WRITING

EXP. NUMBER	EXPERIMENT/SUBJECT		DATE	
NAME		LAB PARTNER	LOCKER/DESK NO.	COURSE & SECTION NO.

SIGNATURE		DATE	WITNESS/TA	DATE

EXP. NUMBER	EXPERIMENT/SUBJECT		DATE	
NAME		LAB PARTNER	LOCKER/DESK NO.	COURSE & SECTION NO.

SIGNATURE	DATE	WITNESS/TA	DATE

EXP. NUMBER	EXPERIMENT/SUBJECT		DATE	
NAME		LAB PARTNER	LOCKER/DESK NO.	COURSE & SECTION NO.

SIGNATURE	DATE	WITNESS/TA		DATE

NOTE: INSERT DIVIDER UNDER COPY SHEET BEFORE WRITING

EXP. NUMBER	EXPERIMENT/SUBJECT		DATE	
NAME		LAB PARTNER	LOCKER/DESK NO.	COURSE & SECTION NO.

SIGNATURE	DATE	WITNESS/TA	DATE

NOTE: INSERT DIVIDER UNDER COPY SHEET BEFORE WRITING

EXP. NUMBER	EXPERIMENT/SUBJECT		DATE	
NAME		LAB PARTNER	LOCKER/DESK NO.	COURSE & SECTION NO.

SIGNATURE	DATE	WITNESS/TA		DATE

EXP. NUMBER	EXPERIMENT/SUBJECT		DATE	
NAME		LAB PARTNER	LOCKER/DESK NO.	COURSE & SECTION NO.

COPY

SIGNATURE	DATE	WITNESS/TA	DATE

EXP. NUMBER	EXPERIMENT/SUBJECT		DATE	
NAME		LAB PARTNER	LOCKER/DESK NO.	COURSE & SECTION NO.

SIGNATURE	DATE	WITNESS/TA	DATE

NOTE: INSERT DIVIDER UNDER COPY SHEET BEFORE WRITING

EXP. NUMBER	EXPERIMENT/SUBJECT		DATE	
NAME		LAB PARTNER	LOCKER/DESK NO.	COURSE & SECTION NO.

COPY

SIGNATURE	DATE	WITNESS/TA	DATE

THE HAYDEN-McNEIL STUDENT LAB NOTEBOOK NOTE: INSERT DIVIDER UNDER COPY SHEET BEFORE WRITING

EXP. NUMBER	EXPERIMENT/SUBJECT		DATE	
NAME		LAB PARTNER	LOCKER/DESK NO.	COURSE & SECTION NO.

SIGNATURE		DATE	WITNESS/TA		DATE

THE HAYDEN-McNEIL STUDENT LAB NOTEBOOK

NOTE: INSERT DIVIDER UNDER COPY SHEET BEFORE WRITING

EXP. NUMBER	EXPERIMENT/SUBJECT		DATE	
NAME		LAB PARTNER	LOCKER/DESK NO.	COURSE & SECTION NO.

SIGNATURE		DATE	WITNESS/TA		DATE

NOTE: INSERT DIVIDER UNDER COPY SHEET BEFORE WRITING

EXP. NUMBER	EXPERIMENT/SUBJECT		DATE	
NAME		LAB PARTNER	LOCKER/DESK NO.	COURSE & SECTION NO.

SIGNATURE	DATE	WITNESS/TA	DATE

THE HAYDEN-McNEIL STUDENT LAB NOTEBOOK

EXP. NUMBER	EXPERIMENT/SUBJECT		DATE	
NAME		LAB PARTNER	LOCKER/DESK NO.	COURSE & SECTION NO.

COPY

SIGNATURE	DATE	WITNESS/TA	DATE

NOTE: INSERT DIVIDER UNDER COPY SHEET BEFORE WRITING

EXP. NUMBER	EXPERIMENT/SUBJECT		DATE	
NAME		LAB PARTNER	LOCKER/DESK NO.	COURSE & SECTION NO.

SIGNATURE	DATE	WITNESS/TA		DATE

EXP. NUMBER	EXPERIMENT/SUBJECT		DATE	
NAME		LAB PARTNER	LOCKER/DESK NO.	COURSE & SECTION NO.

SIGNATURE	DATE	WITNESS/TA	DATE

EXP. NUMBER	EXPERIMENT/SUBJECT		DATE	
NAME		LAB PARTNER	LOCKER/DESK NO.	COURSE & SECTION NO.

SIGNATURE		DATE	WITNESS/TA		DATE

EXP. NUMBER	EXPERIMENT/SUBJECT		DATE	
NAME		LAB PARTNER	LOCKER/DESK NO.	COURSE & SECTION NO.

100

SIGNATURE	DATE	WITNESS/TA	DATE

THE HAYDEN-McNEIL STUDENT LAB NOTEBOOK

NOTE: INSERT DIVIDER UNDER COPY SHEET BEFORE WRITING